1. Preface

Dawn was breaking over the Sea of Japan on September 1, 1983 when a Korean airliner with 269 people disappeared — shot down by a Soviet military jet.

With a flight number evoking James Bond — KAL 007 — the disaster became a Cold War incident, like so many where the sharp edges of nuclear nations snagged and sparked. To this day some of the files remain classified.

KOREAN AIRLINES FLIGHT 007, 01SEP1983

But KAL 007 stands out in history as a tipping point between the frozen madness of mutual obliteration and the warmth of interdependent globalization. Four days later, the United States revealed a secret space program that, it said, could have prevented KAL 007 from straying into enemy airspace.

It was called the Navstar Global Positioning System (GPS). When GPS was fully functioning, President Reagan promised, the world's airliners could find their way with nearly the same accuracy as America's 6,000 nuclear missiles then pointed at the Evil Empire, for which the system had been invented. "The U.S. is prepared to do all it can for this noble aim," Reagan said.

Reagan would never know how much his gesture, which in 1983 was as much propaganda as altruism, would turn out to hasten a technological revolution. The U.S. would eventually build a system considered both "the greatest advance in military navigation and guidance ever fielded," and a mighty electronic glue binding the modern world together.

Fifty years hence historians will place GPS on the list of civilizing inventions, alongside the clock, electricity and the internet —

developments that, while not needed to sustain life, seem indispensable. We couldn't imagine living without them. In fact, without them, civilization as we have come to expect it would sink into chaos.

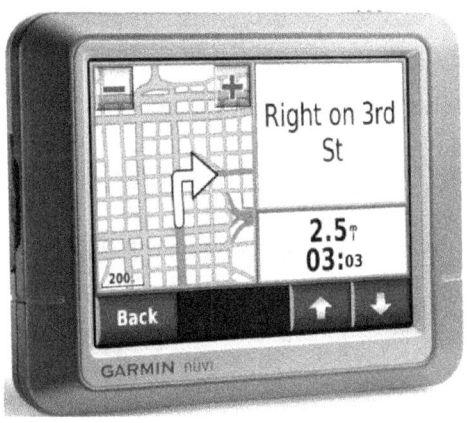

This revolution, still unfolding, has largely been unheralded. We know about palm-sized electronic maps with pleasant voices that guide us turn-by-turn. We hear about dramatic rescues at sea — there have been more than 2,000 lives saved. We may even realize that on computer screens across the world, trucks and ships, sex offenders and grizzly bears move with location precision.

But we likely do not understand the extent to which GPS has stealthily seeped into the world's nervous system, civilian and military, to a point where we are very nearly dependent on it.

Like historic leaps that led mankind to make maps and sail off them, using new-fangled clocks and compasses, GPS created possibilities. By capturing in a little box the spirits of three dimensions, GPS brought touch-screen simplicity to puzzles first posed outside caves: Where am I? What time is it? Where am I going? How do I get there?

By calculating instant answers any place on earth, GPS suddenly made antique the instruments that we'd used for hundreds of years to tell time, find ourselves, and navigate. As if Ptolemy appeared in our palms, we rediscovered fourth-grade geography and found it fascinating and

fundamental, useful and irreplaceable. Simply put, by making time, location and navigation easy, ubiquitous and free, the United States launched a new age.

The GPS Age, now upon us, is not about discovering terra incognita, but rather, about seeing the land in new ways, and relating to earth, sky and each other. To borrow an accolade from the age of discovery, GPS proved to be another "Amplifier of the World."

Today, the U.S. military considers GPS as water — they don't go without it. Proven in the trackless deserts of Iraq, GPS is now fitted on half a million weapons and become so pervasive in central command, "its loss could be catastrophic," according to the Pentagon.

And the same American ingenuity that built the system spawned civilian applications that outnumber the military's 100 to 1.

Twenty-five years ago, as word leaked of a navigating signal from space, garage gadgeteers began wiring boxes that could find themselves in the middle of nowhere. The first sets cost $100,000 and were as large as refrigerators. But once invented, their uses were obvious. Ships and oil rigs were the first adopters. Airplanes were next.

As receiver prices fell 30 percent a year and shrank to fit into cell phones, GPS found use in the fields of science, transportation, construction, search and rescue and recreation. For a world already well mapped, wired and radioed, it was eye-opening how useful location and time could be.

Because of the system's unparalleled delivery of time — nanosecond time — GPS quietly began to undergird the very infrastructure of our modern nation. It synchronizes electric grids and the internet, mobile phones and world banking transactions, including my last $40 ATM withdrawal.

With confidence came reliance — to the point, for example, where the nation's air traffic system, with 50,000 planes in the air any day, will soon depend on it to keep planes apart and landing automatically.

No one has totaled U.S. use, but by one official estimate, 65 percent of Europe's population will rely on space navigation by 2020 as they go about their business and daily lives.

Which is one reason why the European Union, Russia, China, Japan and India launched their own satellite navigation systems to compete with or augment GPS. Economic and military independence, freedom from American cultural and political dominance, and the haunting vulnerability that so much depends on one aging system of 24 satellites, is driving this move.

If, as the U.S. proposes, these systems can work compatibly, space navigation will become even more vital.

From the day in 1962 when a Polaris submarine surfaced somewhere in the Pacific and aimed its nukes with a new secret satellite system, to last Easter Sunday when a dashboard box soothingly guided cousin Tad to grandmother's house somewhere in the suburbs, the invention and growth of GPS inscribes a remarkable arc.

This is that story.

2. In the Beginning...

"Listen," began a prescient newscaster on Friday evening October 4, 1957, "for the sound that forevermore separates the old from the new."

There followed the scratchy beep-beep of Sputnik. All over the world, ham operators, lab technicians, kids and their fathers rigged up receivers. On Monday, at the Johns Hopkins Applied Physics Lab, George Weiffenbach and Bill Guier, two be-speckled junior brainiacs, rigged up a spectroscope, tuned in, and noticed a warble.

It took only minutes to recognize the warble as the Doppler effect of the satellite zooming overhead at 18,000 miles an hour. The beep's highest frequency marked the satellite's closest position to the lab. With scrounged gear and borrowed time on a new Univac "computer," Weiffenbach and Guir were soon able to predict the time and path of Sputnik's 96-minute orbits.

The U.S., meanwhile, scrambled to launch its own satellite. Russia's feat had upped the ante in the Cold War. Both countries now had H-bombs, and the race was on to build missiles to deliver them. Among the projects was the Polaris, to be carried by submarines. The problem was navigation. On a cloudy day, there was no way for the sub to fix its position and aim the missile.

Six months after their Sputnik experiments, Weiffenbach and Guier were summoned by the lab director, Frank McClure. He closed the door.

Are you exaggerating the accuracy of your calculations? He asked.

No, they said.

Well then, couldn't you invert the data? That is, if each orbit could be precisely plotted, using the Doppler effect, couldn't you invert the process, and use the beep from a known orbit to find the position of the lab on the ground?

It was an astoundingly simple idea. Within two days, using the Doppler shift from Russia's newest satellite, Weiffenbach and Guier "found" their location on earth to within one tenth of a mile.

It was the beginning of space-based navigation. Johns Hopkins won a Navy contract to build Doppler-tone satellites to communicate with submarines. It was another innovation born of war, a pattern that stretched back to the Greeks. But not since John Harrison's chronometer solved the "longitude problem," had a single technology created such a leap. The first handmade satellites from Johns Hopkins formed the "Transit" system, and were the first steps toward GPS.

3. All in the Game

The Soviet Union was fully aware of the Doppler effect and had even enlisted amateur tape recordings of Sputnik to plot its position. As the U.S. launched Transit, the Soviets, in 1967, launched Cicada, a virtually identical Doppler system.

Both systems suffered from early satellite mortality, both had blackout periods during the day and the receivers were as large as refrigerators. But they provided the first at-sea navigation with an accuracy of a few hundred yards. While many crusty naval commanders who had come of age with sextants were slow to convince, the Naval Research Laboratory led by Roger Easton forged ahead, eventually launching "Timation," Transit's successor, using orbiting atomic clocks.

The Air Force wanted a system, too, but the Navy's didn't work in three dimensions, and nuclear bombers were roaming the world, including the trackless Arctic. For the Air Force, Ivan Getting of the Aerospace Corporation launched "621B," another atomic clock system that used timed signals rather than the Doppler effect.

By 1972, so typical of U.S. military competition, there were 47 different navigation satellites in orbit or on the drawing boards. While the enemy was supposed to be the Soviet Union, most of the struggle was inside the Pentagon.

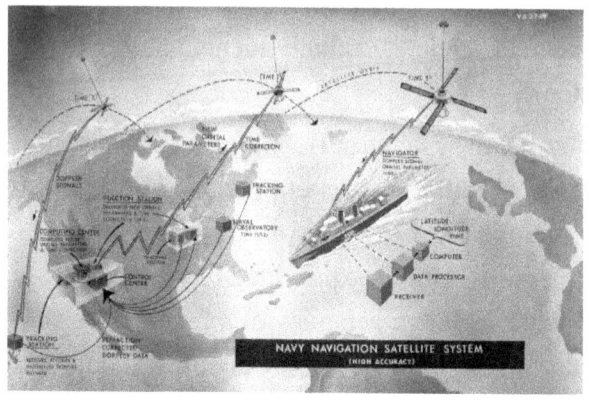

David Packard, cofounder of Hewlett Packard, appointed by Nixon as a new assistant secretary of defense, resolved to end "inter-service bickering" over space navigation by creating a Joint Program Office headed by Col. Bradford Parkinson.

GPS would have died on the table, Parkinson said later, had he not had a chance encounter in Los Angeles with Dr. Malcolm Currie, one of several PhDs appointed by the Nixon Administration to move the military toward high technology.

"This resulted in a remarkable meeting with the number three man in all of the U.S. DOD spending about three hours in a small office with a lowly colonel, talking about engineering, technology, and the wide applications of the proposed system...Without this key decision maker, the Air Force would have killed the program in favor of additional airplanes."

Parkinson, a PhD himself with savvy political skills and the good looks of a California surfer, called a secret Pentagon meeting over Labor Day weekend in 1973. At the meeting, the last barrier to a single, joint GPS system fell, he later confided, when he let a resisting budget wonk suggest a "nice sounding name" — NAVSTAR. It was not an acronym, and GPS, for Global Positioning System, eventually won out, but the ego stroke got the system going.

4. A New North Star

Washington's birthday 1978 dawned mild and foggy north of Surf, California. A February day like so many others, sunny, high 65, this federal holiday would stand out in history because of the Atlas rocket that arched out of Vandenberg Air Force Base at 3:44 p.m. and headed for the North Pole.

The Atlas F, originally built to loft Cold War ICBMs into the Soviet Union, was carrying Navstar 1, a half-ton satellite about the size of a

Volkswagen bug, into a polar orbit. The satellite, blue and gold with photovoltaic wings fifteen-feet long, was the first of a fleet that would prove the accuracy and reliability of GPS.

It was not a cheap undertaking. Each satellite –- there were eleven launched in the next six years — cost $40 million, and each launch cost $25 million in 1970s dollars. Worse, 80 percent of the first atomic clocks installed in the satellites failed.

Congress' watchdog, the Government Accounting Office, warned that the military had vastly underestimated the true cost of a GPS system. At the time, the military was estimating the cost of a "cheap" receiver at $10,000. Because of the cost, the Army only planned to buy 1,600 ground receivers.

Under President Carter, Defense Secretary Harold Brown, a nuclear physicist, kept GPS alive, but barely. Because it wasn't a weapon, GPS had few champions among military brass.

Until 1984 there were only four satellites in orbit, and they were visible and useful only a few hours a day. A plan to use the 1986 space shuttle to launch GPS satellites went up in smoke with Challenger. Not until 1989 did launches resume — just in time for war.

5. Five Bombs in the Same Hole

GPS, the orphan child, was finally adopted by the Air Force for very practical reasons. Pilots needed to know where they were anywhere on the planet, and target their bombs accurately.

The Air Force goal, scrawled on signs in the Pentagon, was to "drop 5 bombs in the same hole." The first test bombs, dropped from a B52 over the Arizona desert, found an astounding accuracy of less than 10 meters. A subsequent cost-benefit analysis showed that GPS would improve the accuracy of air strikes enough to reduce the size of the Air Force bomber fleet by as much as 20 percent. Suddenly, generals in charge found $12 billion to build the system.

Over the next two decades 27 satellites were launched and ground stations were built around the world. To "fly" GPS satellites, the Air Force set up a nondescript windowless building on the plains east of Colorado Springs. Sealed in the building, dressed in trim, blue jumpsuits bearing the breast badge, "Air Force Space Command," seven crewmen monitor the 24 satellites, their orbits and clocks, and keep GPS functioning

"When you first get satellite status, you are very aware of them," said Major Matthew Althouse, the squadron operations chief when I visited. "They tell you, 'One false step and you owe the government $65 million'."

To alleviate boredom, and remind the crewmen that they hold lives in their hands, commanders bring in stories. The success of Norman Schwarzkopf's Desert Storm, the first real GPS war, kept them going for a long time. So, too, the rescue in Bosnia of downed pilot Scott O'Grady who had a GPS in his life vest.

And then there was the father-in-law of one of the satellite crew who was caught in bad weather while flying his private plane. "If he had not had his GPS receiver, he would not be with us," said Althouse. "We get those stories in once every two or three weeks, and we feed those off to

the crew. It's a good motivator for them."

The most important gift from the satellite crew is time. Once a day the atomic clocks in the satellites are synchronized with the GPS master clock, kept in a refrigerated case in a room next door. It is a cesium atomic clock, accurate to one second in a million years, a duplicate of the Naval Observatory clock in Washington. Once reset, the satellites, in turn, reset all the clocks in all GPS units all over the world.

6. Timing is Everything

Behind the magic of GPS is geometry 6,000 years old. The Phoenicians figured out where they were in the Mediterranean by the angles to stars and the first "light houses," pyres erected along the coast. Triangulating at least two light sources (three, ideally), drawing them on a chart, their position lay where the lines crossed.

Measuring and charting angles became an art and science, with increasing accuracy over the centuries. Using a sextant and John Harrison's chronometer, mariners and aviators plotted angles to stars until the 1990s when a new constellation appeared in the firmament.

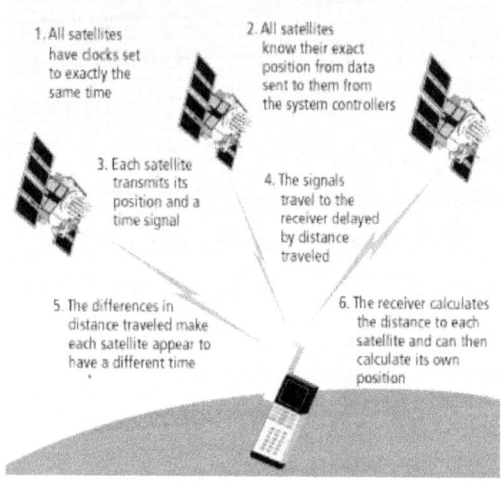

1. All satellites have clocks set to exactly the same time

2. All satellites know their exact position from data sent to them from the system controllers

3. Each satellite transmits its position and a time signal

4. The signals travel to the receiver delayed by distance traveled

5. The differences in distance traveled make each satellite appear to have a different time

6. The receiver calculates the distance to each satellite and can then calculate its own position

"GPS put our own stars up there," said Michael Shaw, the former commander of the Colorado Spring base, who now coordinates military and civilian use. Instead of angles and lines on a chart, GPS plots distances using two clocks, one in each satellite and one in your hand.

Here's how it works: When the satellite beeps, it takes a little time for that signal to go through the atmosphere and thousands of miles to your hand. Because of the two clocks we know the time it left, and the time it arrived. Subtracting, we now know how long it took for the beep to reach earth.

If you multiply that travel time by the beep's speed (roughly the speed of light) you get the distance between the satellite and your hand. If the satellite were overhead, the distance would be about 12,000 miles.

Now, put a second satellite on the horizon where the sun comes up. If we measure a beep from that satellite, we would find it roughly 20,000 miles away.

If you drew an arc from each satellite, one with a 12,000 mile radius, the other 20,000 miles, the arcs intersect. That's where you are on earth.

That is precisely what that GPS in your smartphone does.

In reality, the system is vastly more complicated, involving 24 moving satellites, Einstein's relativity which speeds up the satellite clocks, clock corrections, atmospheric and earthly anomalies (like rain storms, Mt. Everest and global warming), space radiation, and a set of mathematical equations that fill blackboards.

But, essentially, by measuring the distance to three satellites, your hand unit triangulates your position on earth. With four satellites, it can fix your height above sea level. At any given time, at least six and as many as eleven satellites, are visible above the horizon. The more satellite signals your GPS measures, the more precise the position. The standard civilian level today is 15 meters. During military operations, encrypted triangulation can get down to inches.

7. The Spoils of War

Present at birth during the secret Pentagon GPS meeting in 1973 was a rival twin, just as hungry as the military for a GPS signal. Col. Parkinson called it, "the civil problem."

In fact, as early as 1967, military contractors had been allowed to tap into the Navy's Transit system. A number of cargo ships had installed the $100,000 boxes.

So from the beginning, the system designers had reluctantly included a civilian signal on each satellite, alongside the encrypted military tone. This "clear acquisition" code did not require an encrypted receiver on the ground, but neither was it as accurate — it was off by more than 300 feet. The Pentagon also reserved the right to degrade the signal even more by changing the satellite clocks. During war, they didn't want the enemy to use GPS against us.

What they didn't anticipate was a pack of hobbyists, hackers and academics who found a way to tap into the military signal and use it to improve the accuracy of the civilian code to near-military standards.

Four months after the first GPS space shot in 1978, *Navigation*, the journal of the Institute of

Navigation, published all the detail any entrepreneur needed to build a GPS receiver. In garages and attics –- many in what became Silicon Valley — experimenters began wiring boxes to find themselves.

Then came Korean Air 007 from Anchorage to Seoul, shot down on September 1, 1983 by Soviet fighters after crossing over a Soviet peninsula with a military base. Among the dead was a U.S. congressman.

While lambasting Russia, President Reagan issued National Security Decision Directive NSDD 102, (still only partially declassified), offering GPS to all civilian airliners. His offer was accepted by the International Civil Aviation Organization then meeting in Montreal. The fact that this was the plan all along was lost in Reagan's Cold War propaganda coup. Not until 1987, long after civilian entrepreneurs began making and selling the first rudimentary receivers to ships, oil rigs and surveying companies, did Reagan's "pledge" take effect. By then, a number of industries had adopted GPS as their standard.

GPS may go down in history as one government project whose civilian benefits exceed the original military objective. But not by design. The military founders soon found themselves in bed with an insatiable partner who began making demands of its own.

Because the military still runs and pays for GPS, the tension between civil and military users is palpable. To this day, Parkinson writes, "this civil problem is only partially resolved."

8. Charlie Trimble and the Birth of an Industry

Ten years out of Cal Tech, Charlie Trimble was a rising manager at Hewlett Packard, running its integrated circuit R&D shop. He learned on the job, he later told Silicon Valley historian John McLaughlin, because his degree in electrical engineering was useless -- "I had never held a transistor in my hands" –- and because HP then was technology-driven, full of wizards building the future.

One product, for marine navigation, was a LORAN receiver, a technology developed at the end of World War II, and widely used by professional sailors. But HP dropped LORAN as too small a market.

Trimble and three bachelor sailing buddies –- their boat was named "Sybarite" — scraped together $50,000 for "50 boxes of stuff." They set up shop in 1,000-square feet of space over the old Los Altos theater on Main Street.

The year was 1978, the same year the U.S. launched its first GPS satellite. HP had been tinkering with GPS, too, but by 1982 sold that technology to Trimble, also. Using moonlighting brain power from Silicon Valley, Trimble produced in two years his first GPS receiver for offshore oil-drilling platforms. It was big, slow and expensive, and could only be used a few hours a day because of the handful of satellites, but it revolutionized the way oil companies marked oil deposits and placed their rigs in the trackless Gulf of Mexico.

By 1988, long before the GPS system was officially operational, Trimble had 700 patents and had become the world's GPS leader. The company's goal, Trimble said, was to make its customers money. Trimble did that in oil, surveying, mapping, agriculture, aviation, earth-moving, shipping and fleet management. By 2007 the company sold 500 products and had annual revenues of $775 million.

9. Mailing Cookies — and a GPS

When Saddam Hussein invaded Kuwait, August 2, 1990, the U.S. military found itself in a desert without a camel. With outdated maps, dust storms and a terrain with no features, troops suddenly discovered the value of GPS.

The Defense Department distributed all it had, an inventory that included 900 seventeen-pound backpack units made by Rockwell, and several hundred Trimble Trimpacks, rugged, 3-pound, $3,500 boxes as big as car radios. But as Operation Desert Storm loomed, troops and

commanders begged for more. The Army placed

an emergency order for 7,400 more Trimpacks and 1,000 Magellan civilian units.

`Fewer than half were delivered by the time the U.S. attack began six months later. During the fighting, troops got lost in the desert, friendly fire killed 36 soldiers, and half the operations were at night.

To meet the need, hundreds of soldiers received Christmas gifts of GPS units from parents and wives purchased from retail stores in the states. They duct-taped them to dashboards of tanks, jeeps and helicopters. Because so many civilian units were being used, the Air Force decided not to degrade the civilian signal.

On the first day of the war, in one of the most awesome displays of the technology, thirteen B52s equipped with GPS roared 400 feet over the desert to strike Hussein's air bases. The plan was to leave their radar off to avoid detection, and then turn it on for final targeting. But bombardiers found they didn't need it. Using GPS coordinates their strikes landed within 90 feet of each target, devastating the Republican Guard.

Elsewhere, the first "smart bombs" ever guided by GPS were dropped. Commanders also found that helicopters could maneuver in dust storms, casualties and enemy lines could be accurately marked, and artillery could target while on the move. Last but not least, Norman Schwarzkopf's famous "left hook" offensive through uncharted desert to surround the Iraq front was made possible by GPS navigation.

The Gulf War reversed the reluctance to invest in GPS equipment, according to *Precision Revolution*, a history of military GPS use. "NAVSTAR's applications were so all-encompassing that it is likely the most important improvement in tactical command and control since the radio."

Today, GPS is so thoroughly woven into America's war plans that, according to one Air Force Commander: "GPS is like water. Combat forces don't go anywhere without it now."

10. Billions and Billions

Olathe, Kansas sits on the prairie below a great bend in the Missouri River 250 miles west of its confluence with the Mississippi. Coming upstream from St. Louis, the river bends northwest here and heads into Lewis and Clark territory. But pioneers looking for a faster route west, struck out across land and right through Olathe, (Oh-LAY-thuh).

Years later, perhaps because of its endless, treeless terrain, Kansas birthed several aviation pioneers, including Boeing, Learjet and Beechcraft, and an unusual number of Kansas boys dreamed of being pilots. Ed King of Olathe was one, and after getting his license, set out making a better radio for his plane. In an old farmhouse not far from the Santa Fe Trail, King Radio became world famous in avionics.

Two of King's engineers, Gary Burrell of Olathe and Min Kao of Taiwan, created the first GPS navigator certified for airplanes by the FAA. But after King was sold to a multinational corporation, Burrell left to become a minister. Over dinner Kao convinced him to start a GPS company, whose name would combine theirs: Gary and Min = Garmin. They raised $4 million in Taiwan and set up shop in Olathe.

Their first product, in 1990, was the $2,500 "GPS 100" for recreational boaters. They sold 5,000 units. Today, there are many GPS brands, but Garmin is king with 19 million units sold, a 60 percent market share, 4,500 employees and sales approaching $2 billion. When the company went public in 2000, Burrell and Kao instantly joined Forbes' billionaires list.

Garmin is a prime example of a non-government GPS customer whose very life depends on the "unresolved" issue of civilian-military signals. As the company starkly puts it in its annual report:

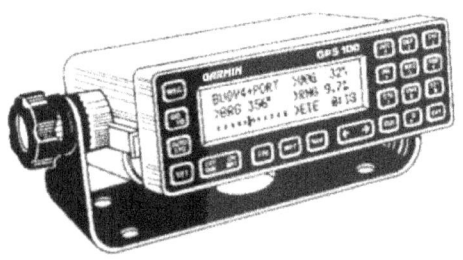

"Our GPS products depend upon satellites maintained by the United States Department of Defense. If a significant number of these satellites become inoperable, unavailable or are not replaced or if the policies of the U.S. government for the use of the Global Positioning System without charge are changed or if there is interference with GPS signals, our business will suffer."

11. New Kids on the Block

The 1990s were GPS's golden years. The system was declared fully operational in 1994. The price of receivers dropped from thousands of dollars to under $100 (Magellan 1997).

GPS seeped into unimagined applications. It became the world's clock, with the internet, power grids, banking transactions and the military relying on it for exact, synchronizing time.

Geodesy, the ancient science of the shape of the earth, was rewritten with precision GPS instruments –- in fact it was GPS's pioneering satellites that first proved that the earth wasn't exactly round. Today, with augmented precision of 1-3 centimeters, GPS instruments are re-measuring earth from pole to pole.

The world's airspace, which since World War II has relied on radar to keep airplanes apart, is adopting a GPS system to separate and guide planes from takeoff to landing.

With an international economic impact approaching $20 billion a year, growing at nearly 30 percent a year and all dependent on GPS, the Clinton Administration decided to end the military's longstanding threat of degraded civilian signals in times of conflict. On May 2, 2000, the Department of Defense "discontinued" the policy. Four years later the Bush administration said the U.S. "does not intend" to mess with the signal again. The world could count on 10 meters or less accuracy, anywhere, anytime.

The pledge came not a moment too soon. America's primacy faced competition from the European Union, the Soviet Union and China, each of whom promised their own satellite navigation system, with stronger signals, and exclusive civilian use. Japan and India also announced regional satellite augmentation.

This new competition reflected both growing suspicion of, and annoyance with, U.S. foreign policy, and a worldwide awakening to the power and utility of space navigation. China, where most GPS receivers are made, was no longer comfortable keeping its bursting economy in the hands of the Pentagon.

"In a few years business without a navigation signal will become inconceivable," a Russian analyst told the New York Times.

It also became apparent that the U.S. had been resting on its laurels. The GPS system was beyond its design life -– one study called it "frail" — and upgrades, including replacement satellites, a stronger signal, and better accuracy, had been delayed for years by budget cuts. Even military upgrades are behind schedule. In Iraq, the U.S. learned that insurgents using a $10 box of electronics could jam the GPS signal over an area the size of Baghdad. The study summed up the sorry status of GPS with its title: "For Want of a Nail."

Europe's GALILEO system promised to deliver a strong, and more accurate, civilian signal -– guaranteed in writing — and charge for it. With the specter of losing its world leadership, and business, the U.S. scrambled to make sure signals from GALILEO and GPS were compatible.

Russia pledged to have its GLOSNASS system remodeled and working by the end of 2007. China's BAIDU, the Chinese word for the Big Dipper, launched its first satellites in orbit too. Either one could be used by America's enemies, including Al Qaeda. Their plans, in a sense, eerily echoed space navigation's roots in the Cold War. The competition of ideologies had become a competition of economies and ideas.

12. Where Are We Now?

Fifteen years ago, when there were one million GPS units in the world, the phone rang at the Jefferson County sheriff's office outside of Denver.

"Where am I?" the caller asked.

He had been hiking in a wilderness area of the Rocky Mountain Front Range, carrying a cell phone and a GPS. But he had no map — no way to apply his whereabouts, read as a strange string of numbers denoting latitude and longitude. A rescue unit pulled out a map, plotted his place, and told him he was west of Evergreen.

Had that hiker gotten lost today, there's a good chance he could have found his way home

using his smart phone, with both cellular and

satellite signals, and a map pinpointing his location as he walked.

In the years since that call, GPS has become a utility, not unlike electricity, embedded and ubiquitous in so many devices and systems that it is taken for granted.

Today innovations come so fast that it is hard to imagine a GPS use that isn't already on a drawing board. Google, for example, which has made GPS mapping and location free to anyone with an internet connection, is testing a GPS guided car, a system that could dramatically reduce one million annual vehicular deaths worldwide.

The dark side of tracking is spying.

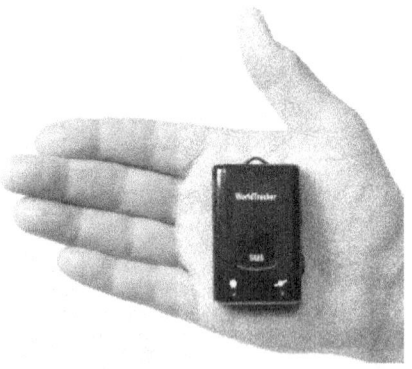

That same system could record one's movements, ending the sense of the "open road."

When a Florida company created a GPS implant about the size of a pacemaker, the outcry shelved its release -– except for ongoing tests by a very interested Justice Department.

But GPS also creates new social space that shrinks the distance between us. Finding friends nearby on your mobile phone, or a late errant teenager on your PC are common applications, harbingers of even more intimate connectedness made possible by technology.

With GPS installed in mobile phones, "location awareness" becomes an unlimited marketplace. Already you can ask your phone to find a parking space in New York City, an art museum in Paris, or the nearest and cheapest gasoline.

New GPS satellites, with new, stronger signals, promise greater accuracy and fewer blank spots on the earth's surface. With accuracies as small as a dime, engineers will be able to watch the movements of earth plates, suspension bridges, and space stations.

But what becomes of "getting lost," of wilderness, and of exploration, if every inch of the globe is mapped and described to us in endless streams? Is knowing always a good thing?

13. One Week in the Life of GPS

At 5:15 p.m. on June 7, 2006 a U.S. jet dropped two bombs on a cinderblock house and killed the most wanted man in Iraq –- Abu Musab Al-Zarqawi, the al Qaeda leader responsible for hundreds of bombings, kidnappings and beheadings. The bombs were "smart," one of them a 500-pound blockbuster fitted with an $18,000 tail kit that contained in its electronic brain the precise location of Zarqawi's safehouse. As the bomb fell, signals from GPS satellites steered it to within a few feet of the target.

One week later and six thousand miles to the west, aircraft raced toward three sailboats sinking in Tropical Storm Alberto south of Nova Scotia. Using coordinates from a $150 GPS receiver the size of a cell phone, radioed from sailor Bob Dwyer aboard *Dad's Dream*, who was bailing water with a bucket and watching cracks in both his hull and his wife's vertebrae, they rescued four crew on two boats.

Abu Musab al-Zarqawi

In the space of this one week GPS killed and saved lives. In Iraq and on the high seas, in between and around the world, GPS enabled a bewildering number of military operations, rescues, industries, hobbies, studies and adventures. Here is an actual accounting of those seven days:

On the day Zarqawi was killed, Bill Adams (not his real name), a pilot flying a corporate jet from LaGuardia to Atlanta, was cleared to land using nothing but GPS guidance from an $18,000 box measuring 4-by-6-by-12 inches. On that day, the FAA added 55 airports, including Atlanta, LaGuardia, Key West, Colorado Springs and Laramie to a list of 300 airports where enhanced GPS signals create landing flight paths accurate to two meters horizontally and vertically.

In Newton, N.J., that morning, Thomas Ziniewicz was sitting in jail after his GPS ankle bracelet showed he'd spent 21 previous days at his girlfriend's apartment without notifying authorities. Under Megan's Law, more than 200 New Jersey sex offenders must wear the bracelets, which cost the state $5 to $10 a day per man.

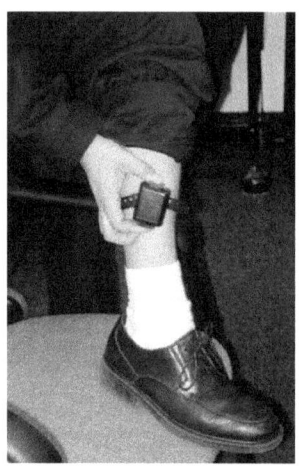

Ziniewicz's trysts had begun a few days after Disney Mobile unveiled a combination cell phone and GPS to keep track of children. Civil libertarians and not a few kids frowned upon the concept.

In Lawrence, Kansas that same day, Tim Hibbard wrapped a GPS phone in a box, handed it to a UPS clerk, and invited the world to watch as his Web site tracked the phone, minute-by-minute to its destination, a friend's house in Louisville, Ky. The battery died before delivery, but it proved a good publicity stunt for Hibbard's employer, Engraph, which sells real-time GPS-tracking system for motor vehicles.

Installed in virtually every name-brand delivery truck in the U.S., including those owned by Wal Mart, Sears and FedX, GPS systems keep track of trucks, improve route efficiency, spy on driver speeds and stops, and allow supervisors to open doors and turn off motors with a remote command.

In Montgomery, Alabama, John Taylor, a retired banker, walked onto the Robert Trent Jones golf course and shot an 85, thanks in part to his $390 GPS SkyCaddie that measured the distance from his ball to the tee, to within 5 yards.

At the Black Thunder mine in Wyoming, a behemoth D11 Caterpillar pushed dirt around a black coal seam while the operator watched a computer screen displaying a little bulldozer plowing through virtual dirt, red for what needed moving and blue for what needed covering. Guided by GPS with an accuracy of three centimeters, the operator's job was to "push the red to the blue," said surveyor Jim Long. "It is really slick. They don't even have to look out the window, except to keep from running over somebody." When first installed, the systems increased productivity 30 percent, which in a mine that produces 47 million tons of coal a year, is a pile of money.

On the island of Sumatra in Southeast Asia, 38 GPS stations were alert for a somewhat smaller earth movement –- 7 millimeters, to be exact. That was the amount they measured the day after Christmas 2004, a lurch that presaged the tsunami that killed 240,000 people. Scientists hope to turn the GPS system into a forecasting and warning tool for future tsunamis.

On a slightly faster track, Reuben Halper was unwinding from the Memorial Day Indy 500, where his GPS-enabled Sportvision had, for the first time on ESPN tracked the 200 mph cars within two centimeters of their real-time positions. With other whiz-bang special effects, Sportvision had made the long car race actually worth watching.

Off New Jersey, meanwhile, hurricane hunters were dropping GPS dropsondes into the eyewall of storm Alberto in an attempt to measure its strength and direction. Their information changed the forecast to veer off the East Coast and into the path of three sailboats off Nova Scotia.

On a more bucolic coast, an elephant seal surfaced off Pescadero, California, a GPS headset on its head. After dives to 1,000 feet a radio set beamed the seals' position along with a three-dimensional reconstruction of its underwater habits. In the wilds of Alberta, at the same time, a Canadian grizzly wearing a GPS camera strolled about providing researchers with a bear's eye view of its habitat.

In San Francisco, tourists were renting GPS-guided audio cars for tours of the Embarcadero. The talking scooters gave directions and an insider's view of the city's sites.

Meanwhile, in the dark of noon in Morristown, New Jersey, John Hess and guide-dog Willie were exploring new neighborhoods with Trekker, a GPS computer that he wore on his hip, its synthetic voice calling out street names. "I hated walking around Morristown," said Hess, who is legally blind. "Now, I've gotten adventurous."

As the week came to a close, Brian Ruecki watched corn grow in a field outside Green Bay, Wisconsin. Using GPS to plot planting, just as farmers do in tractors, a technology that has vastly improved farm efficiency, Ruecki had carved in 40-foot letters: "STACY: WILL YOU MARRY ME?"

By late June, when the corn was waist high, the letters stood out clearly. Three weeks later Ruecki lured his girlfriend, Stacy Martin, into a plane and flew over the field.

She said yes.

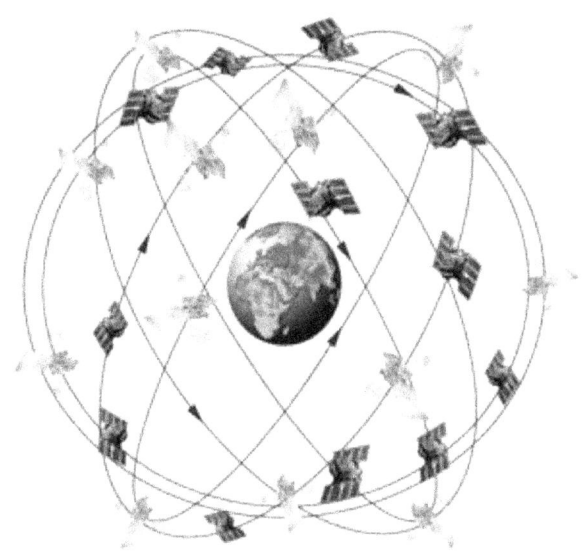

14. A Note From the Author

I have studied and used GPS as a navigator, reporter, and entrepreneur. I own half a dozen GPS units, and have used them to cross the

Atlantic Ocean, navigate through ship channels in pea-soup fog, find dinner appointments in strange towns, and create audio tours triggered by GPS signals.

I still have my first one, a Garmin "GPS II" made in 1995, which cost $200. Rugged and rudimentary, about the size of an early mobile phone, it never fails to astound me with its magic.

Just now, I popped in four AA batteries, turned it on, and read the message, "Searching the Sky." Within seconds, I could read my position, altitude and the time.

After punching in the coordinates for Paris and pressing, "Go To," a little compass-like arrow swung northwest, and promised that Paris lay 3,993 miles away. Propped in my windshield as I ran errands, it predicted that at my pace of 40 mph I would arrive in Paris in 99 hours, just in time for a late lunch four days hence.

I have written about GPS for the *New York Times*, the *Denver Post* and *Cruising World*. These pieces took me from the Air Force's GPS satellite command center in Colorado Springs, to coal mines in Montana and to shrimp boats in the Gulf of Mexico whose captains marked shrimp lodes as if they were "Xs" on a treasure map.

In 2002, I crossed the Atlantic in my 35-foot yawl, *Ranger*. When people ask how long it took, I reply: "Thirty days –- same as Columbus. Except I knew where I was, plus or minus 15 meters." Although I carried a sextant, I relied on three GPS units. I used one plugged into my laptop to make a tricky nighttime approach to Bermuda. On the screen, I watched a virtual *Ranger* enter the narrow channel of St. George's Harbor and come to rest at anchor.

When I flew home I rented a car with a dashboard navigator and drove into New York City, stunned by its accurate, turn-by-turn guidance. This experience prompted me to start a business, IntelliTours, which used GPS signals to trigger automatic audio tours.

My newest GPS, a 2009 Garmin Nuvi, is the size of a pack of cards, cost $150 and contains the entire U.S. road system. Punching in an address, my "car" suddenly appears on a colorful map while "Jill," a robotic voice, leads the way.

The years between my first and sixth GPS units coincided with the official life of GPS. Declared fully operational in 1994, a mere adolescent at 13 GPS had become a worldwide asset that changed the way we see the earth and each other.

I wrote this primer to fill a hole in the literature of GPS, which focused on the "how-to." What's important to me are stories: the system's history told by the people who created it; the knowledge and wealth GPS creates; the tensions between knowing and spying, civilian and military; and the dichotomy between life-saving and death-dealing from the same heavenly messenger.

Photo and graphics

KAL 007 map (CIA)
Nuvi (Garmin)
Ptolemy (Wikipedia)
Sputnik newsreel (Johns Hopkins University -
Applied Physics Laboratory)
Globe with Weiffenbach, McClure, Guier (Johns
Hopkins/APL)
Transit system (Johns Hopkins/APL)
Brad Parkinson (Aero Corp.)
GPS launch 2004 (U.S. Air Force)
GPS atomic clock (U.S. Air Force)
How it works graphic (Aero Corp)
Soldiers with GPS (U.S. Army)
Desert Storm GPS (U.S. Air Force)
Smart bomb (U.S. Air Force)
Gary Burrell and Min Kao (Garmin)
GPS 100 (Garmin)
Magellan Pioneer (Magellan)
Galileo launch (European Union
Hiker with Etrex (Garmin)
GPS Tracker (World Tracker)
Al Zarqawi (CNN)
Ankle GPS (State of Massachusetts)
Elephant seal (University of California)
GPS constellation (U.S. Air Force)
Jim Carrier (Chris Rossin and Roberto Cappi)